GRAVITY WILL ALWAYS PULL YOU DOWN... UNLESS YOU'RE AN ASTRONAUT

A child's introduction to gravity

By Lawrence Martin
Illustrations by Rebecca Weisenhoff

Gravity Will Always Pull You Down… Unless You're an Astronaut
By Lawrence Martin
Illustrations by Rebecca Weisenhoff

Lakeside Press in Association with AimHi Press
The Villages and Orlando, Florida
www.AimHiPress.com

Copyright 2016, Lawrence Martin

Summary: Gravity Will Always Pull You Down is a fully illustrated interactive look at gravity for elementary school level children. Perfect for families and schools teaching about gravity and space.

Picture Credits:
Cover: Astronauts working on the International Space Station. Public Domain: http://spaceflight.nasa.gov/gallery/images/shuttle/sts-116/html/s116e05983.html Pages 1-8, 16-17, 19, 21-24, 28-34: Rebecca Weisenhoff
Page 18 – Public Domain: https://commons.wikimedia.org/w/index.php?curid=48209 Page 9 – Joanna Martin
Pages 10-15, 1, 20, 25-27 licensed under Shutterstock.com
Page 37: Public Domain: https://commons.wikimedia.org/wiki/File:Prinicipia-title.png

All rights reserved. No part of the material protected by this copyright may be reproduced or utilized in any form by any means without permission in writing from the copyright owner.

For Eli, Iris, Maya and Asher. They are my gravity.

You jump up, and what happens?

You come back down! Why?
Because of GRAVITY.

Gravity makes you come back to the ground. If there was no gravity you would keep going up and up, into the sky!

Do you like to throw a ball? Without gravity, it would never come down! Gravity makes the ball come back to earth.

Gravity is the reason jumpers and balls come back to earth. No one thought much about gravity until Sir Isaac Newton, an English Scientist, wrote about it over 300 years ago. Legend has it that one day he saw an apple drop from a tree, and that made him think about gravity. We don't know if this actually happened, but Newton was the first scientist to write about gravity. His book about gravity was published in 1686. It is full of math. Newton was smart enough to be called a genius.

So where does gravity come from? No one really knows, but gravity is part of all matter so everything you can see and touch has gravity. Gravity is what is called a force. You jump up and then come back down because of the *force of gravity*.

Now here's something important to keep in mind. The bigger the object, the more gravity it has. Or, to be more precise, the more mass an object has, the more gravity it exerts. "Mass" means solid stuff. Thus if you have a balloon filled with air and a rock of the same size, the rock would have more mass and so would exert more gravity. What's the biggest object of all (the one with the greatest mass) that we come into contact with every day? A car? A house? A skyscraper? A mountain?

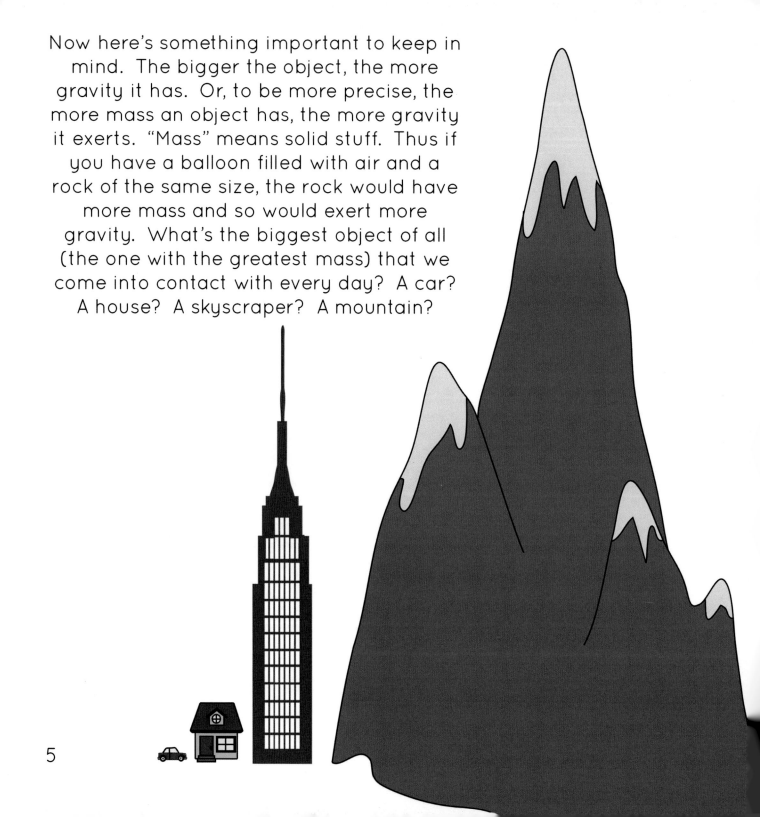

The answer is – the Earth! Earth is our planet and is by far the biggest, most massive thing we come into contact with every day – unless you're an astronaut and live in outer space.

And since earth is so large compared to anything on its surface, you feel the effects of earth's gravity every time you lie down, sit, walk, play ball, or do any other activity. Gravity is always there, pulling on you.

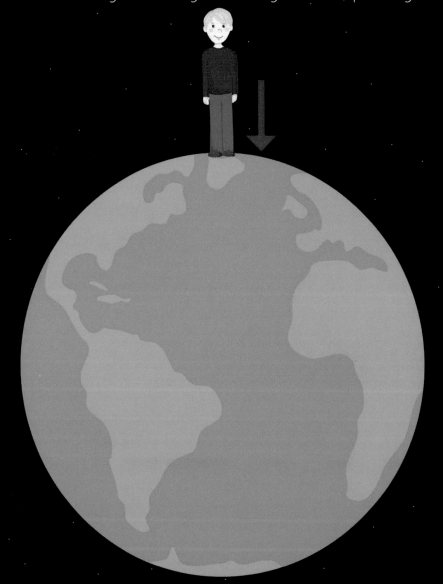

So when you jump up, it is earth's gravity that brings you back down. And that's a good thing. Without gravity, you would just float away!

But when is gravity not good for you? If you fall from a tree.

But you know that!

And if you like to rock climb, it is always a good idea to prepare for a possible fall. Either use a harness to keep you from falling or, as with this brave boy, have a nice padded floor below. Remember, gravity is always there. Be prepared!

Sometimes gravity is good, sometimes not so good. If you're careful, you'll know when gravity can hurt you. But as long as you stay on earth, you will always feel the effects of gravity.

Birds and butterflies and airplanes can overcome gravity by using their wings. A bird flaps its wings to create enough pressure to stay up in the air. But eventually the bird has to come back down; it can't stay up forever.

The same thing is true of the butterfly. By flapping its wings the butterfly can overcome gravity for a short while.

An airplane overcomes gravity by using powerful engines to move it forward through the air. This movement creates a lower air pressure over the wings, so it stays up in the air. But without fuel to power the engines, the plane would fall – due to gravity. Of course, airplane pilots always land the plane well before it runs out of fuel.

So birds and butterflies and airplanes can fly, but the force of gravity means they will always have to come back down to earth.
There is one way to overcome the effect of gravity. Fly in outer space, where the astronauts go!

The fastest way to get into outer space is with a powerful rocket ship. A rocket ship needs a tremendous amount of energy to leave earth and reach outer space. This picture shows a rocket ship carrying the Space Shuttle into outer space, where it will go into orbit around the earth.

Once close to outer space, the rocket releases the space shuttle and falls back to earth – due to gravity. The space shuttle continues under its own power until it reaches orbital height, about 200-300 miles above the earth.

There is still lots of gravity in the area of outer space close to the earth. How much? You may be surprised at the answer.

The United States Space Shuttle program ended in 2011. Now astronauts orbit earth in the International Space Station, about 250 miles above the earth. Just as in the Space Shuttle, astronauts are weightless in the Space Station. You can see them float in internet videos.
Why are they weightless?

Many people think it's because there's no gravity in outer space, but that's not true. At 250 miles up there is still about 89% of earth's gravity. Yet the astronauts are weightless because of what scientists call *microgravity*. Microgravity occurs when the speed and direction of the Space Station balance out the pull of earth's gravity.

Microgravity

To achieve microgravity - where the astronauts are weightless - the International Space Station travels at 17,500 miles an hour at a distance of about 250 miles from the earth. The speed and direction of the Space Station balance out the pull of earth's gravity, so everything inside floats if not tied down.

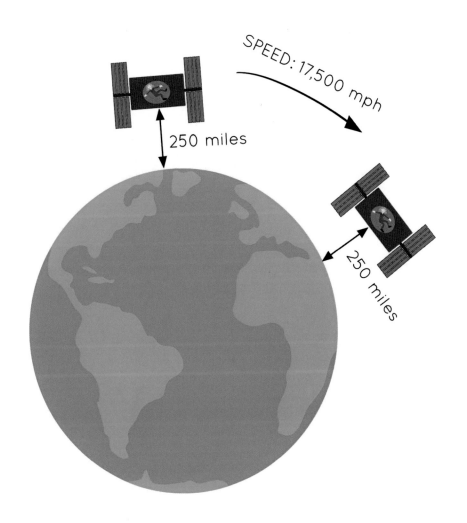

While outside the spaceship, astronauts travel at the same speed and are weightless. This photo, from 1984, shows the very first test where an astronaut was not connected to the spaceship. So-called untethered spacewalks were tested during the Space Shuttle era using special maneuvering equipment. When astronauts work outside the International Space Station they are always connected in some way, via either a tether or mechanical arm.

Being weightless could be fun – for a while. But imagine if every time you sat down to eat (buckled in, of course), your food floated around you. Then it's not fun. On earth, gravity keeps things – and people – from floating away.

What if you traveled to the moon? The moon is big, but not as big as earth. It would take several moons to equal the size of the earth.

MOON

EARTH

The moon - like all objects - has gravity, but because it's much smaller than the earth, it has much less gravity. Since the amount of gravity determines how much you weigh, on the moon you would weigh less than on earth. How much less?

What you weigh on EARTH	What you would weigh on the MOON
40 lbs.	7 lbs.
60 lbs.	10 lbs.
80 lbs.	13 lbs.
100 lbs.	17 lbs.

However, on the moon you would still have the same strong muscles and bones you have on earth, so you could jump much higher — even with your heavy space suit! And you could throw a ball much further. Why? Because the moon has so much less gravity than the earth!

Gravity is important not just because it keeps us attached to earth, but also because it keeps the planets and the moon from floating away. The moon goes around the earth in almost a circle, kept in place by the pull of earth's gravity. Without earth's gravity, the moon would drift off into deep space.

Moon goes around the earth, kept in place by the pull of earth's gravity.

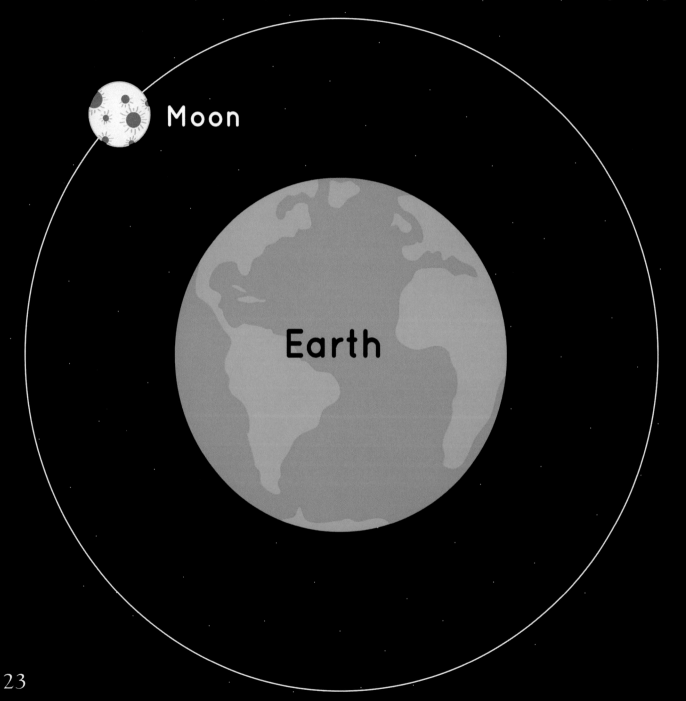

But the moon also has gravity, though not as much as the earth. The moon's own gravity "pulls" on the earth's oceans, causing the tides. High tide is when the moon's gravity pulls the strongest and low tide when it pulls the weakest.

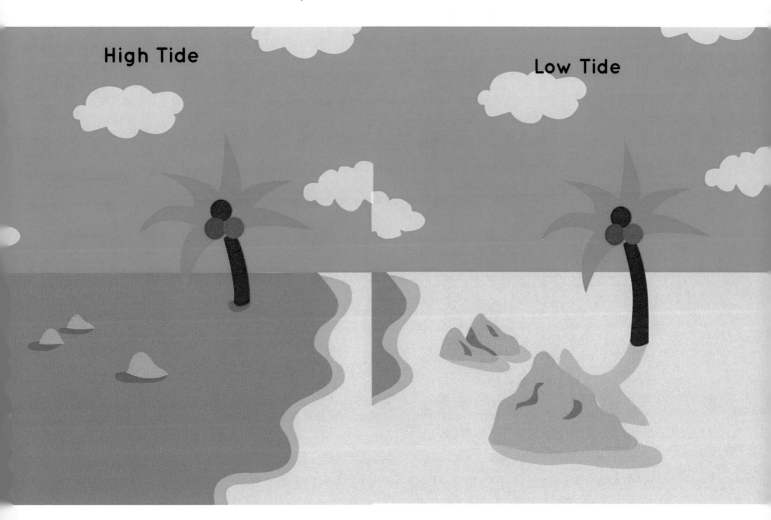

If there was no moon there would be no moon gravity, and we would see very little change in the tides.

Tides Caused by Gravitational Force of the Moon

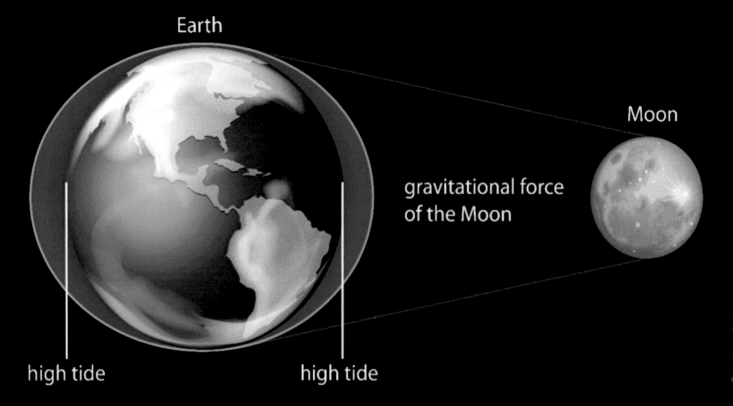

The side of the earth closest to the moon and the opposite side of the earth at that point, have the highest tides. The ocean closest to the moon bulges out in the direction of the moon. Another bulge occurs on the opposite side of the earth, because the earth is also being pulled toward the moon and away from the water on the far side. Finally, since the earth rotates while all this is happening, there are two high and low tides each 24-hour period.

Now for something bigger and more massive. Remember, the more massive an object is, the more gravity it has. And what's bigger and more massive than the moon and the earth? The Sun!

The sun makes daylight. During the day, you can see the sun whenever it's not hidden behind clouds (you should never look at it directly, though).

The sun is bigger than the earth and the earth is bigger than the moon.

The sun is much bigger (and more massive) than the earth, and has so much gravity, that it makes the earth rotate around it. Thus the moon rotates around the earth and the earth rotates around the sun. All due to gravity!

The sun is at the center of our solar system. It is so big and has so much gravity, that it makes all the planets, including earth, rotate around it. Without gravity, the earth – and all the other planets of our solar system – would become totally dark and fly off into space! Can you name all 9 planets? Pluto, the smallest, is now called a "dwarf planet."

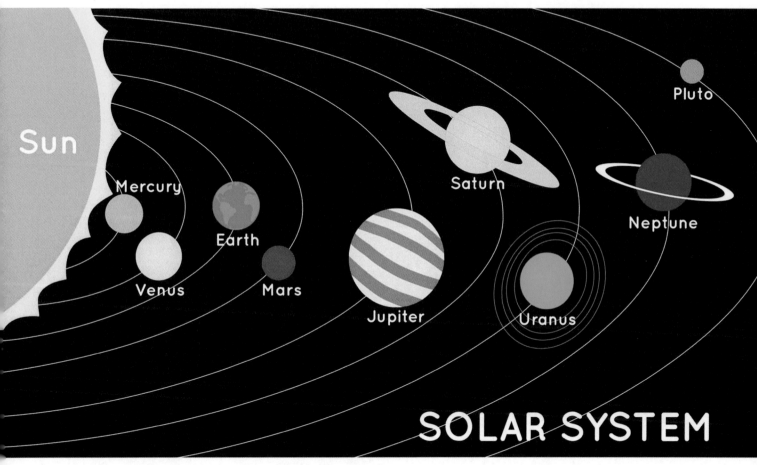

So the sun's immense gravity keeps all the planets traveling around the sun. And earth's gravity keeps our moon traveling around the earth. And the moon's gravity makes our oceans rise and fall. Gravity is a powerful force. Earth has just one moon but some planets have more than one moon. Mars has two moons, called Phobos and Deimos. They rotate around Mars because of Mars' gravity.

And the same thing is true of Saturn's rings. Those rings are made up of millions of rocks. They all rotate around Saturn because of Saturn's gravity. If Saturn were to somehow disappear, all the rocks in the rings would fly off into space.

Saturn's rings are made up of millions of individual rocks.

Will you ever go into space where you can live in a weightless environment?

Possibly. One day people will fly to the moon or to Mars just like we fly in airplanes now. We don't notice anything different about our weight in airplanes because they still fly close to earth, and at much slower speeds than the Space Shuttle and Space Station.

But in a future spaceship, far from earth, you may experience weightlessness.

Whether or not you are weightless will depend on how fast the spaceship is going and its distance from the earth, moon, sun and other planets. If you are weightless, it will be more important than ever to "strap on your seat belt." That is, unless you want to float around like an astronaut.

So remember, unless you're in outer space, gravity is always going to pull you, and everything else, back to the ground. That's both good and bad. Drop a dish on a hard floor and it will break (not good).

Throw a ball in the air and it will come back down (good). On earth there will always be gravity. And that's mostly a good thing. It keeps you from floating into the sky when you jump. On earth you will always feel the pull of gravity. And that's mostly a good thing.

Without gravity milk would float out of an open glass.

And when you down to go to sleep, gravity keeps you in bed without straps that the astronauts have to use. Gravity is good when you are sleepy.

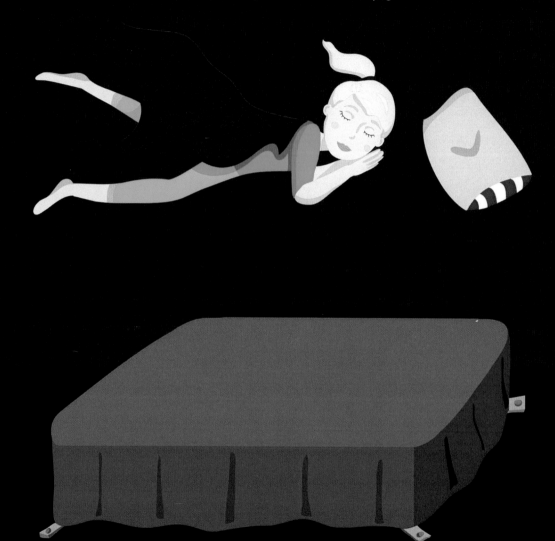

If one day you travel into space like the astronauts, it will be fun to see what life is like without gravity.
But after a while, I think you will miss having gravity.

A Short Quiz About Gravity

1. Below is a picture of all the planets that rotate around the sun. On which planet would you weigh the most?

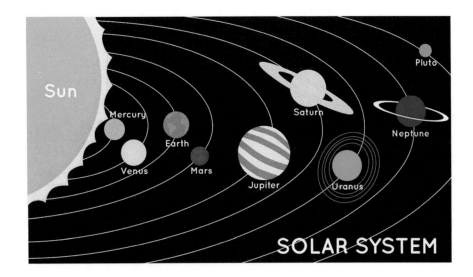

a) Mars
b) Venus
c) Jupiter

Answer: c) Jupiter. Jupiter is the biggest planet, and so has the most gravity. The more gravity there is, the more you weigh.

2. On which planet would you weigh the least?

a) Pluto
b) Neptune
c) Saturn

Answer: a) Pluto. Pluto is the smallest planet, and so has the least amount of gravity. Pluto is now called a "dwarf planet" because it's so small.

3. The International Space Station orbits the earth at about 250 miles up. Astronauts are weightless because:

a)   There is no gravity that far from the earth.
b)   There is microgravity in orbit.
c)   The Space Station travels in a vacuum.

   Answer: b) Microgravity occurs when the speed and direction of the Space Station balance out the pull of earth's gravity.

4. If you weigh 60 lbs. on earth, how much would you weigh on the moon?

a)   5 lbs.
b)   10 lbs.
c)   15 lbs.

   Answer: b) 10 lbs. See page 21.

5. Gravity is present in every object, big or small.

a)   True
b)   False

   Answer: a) True. Because the earth is so much bigger than anything it contains, we only feel the effect of earth's gravity.

6. The bigger the object, the more gravity it has.
a)   True
b)   False

   Answer: a) True

7. If you fly in an airplane you will still have the effect of gravity.

a) True
b) False

   Answer: a) True. The airplane flies close to earth so there is just about the same amount of gravity as on the ground.

8. The object with the most gravity in our solar system is

a) The sun
b) The earth
c) Jupiter

   Answer: a) The sun, because it's the biggest (most massive) thing in our solar system. The sun's gravity is so great that it makes all the planets go around it. Without the sun the planets would all fly off into space.

9. What scientist first discovered gravity and wrote about his discovery?

a) Einstein
b) Newton
c) Galileo

   Answer b): Isaac Newton, over 300 years ago. He was from England, and became very famous after writing about gravity.

10. How would you describe gravity?

a) Something you can avoid if you want to
b) A force that is always present
c) A force only found on earth

   Answer: b) A force that is always present. Even in outer space!

PHILOSOPHIÆ NATURALIS PRINCIPIA MATHEMATICA.

Autore JS. NEWTON, Trin. Coll. Cantab. Soc. Mathefeos Profeffore Lucafiano, & Societatis Regalis Sodali.

IMPRIMATUR·
S. PEPYS, Reg. Soc. PRÆSES.
Julii 5. 1686.

LONDINI,

Juffu Societatis Regiæ ac Typis Josephi Streater. Proftat apud plures Bibliopolas. Anno MDCLXXXVII.

Isaac Newton's book about gravity - published in 1686!

Lawrence Martin is a retired physician who has written medical
books on oxygen, pulmonary physiology and scuba diving.
He also writes about science for children.
His email is drlarry437@gmail.com.

Rebecca Weisenhoff is an artist-illustrator. A graduate of American
University, she is a freelance designer and illustrator with
a passion for children's book illustrations.
Her email is rlweisenhoff@gmail.com.

Visit AimHiPress.com for more books and other products from AimHi Press, NCG Key, and the rest of the Newhouse Creative Group family!

Made in United States
Orlando, FL
23 March 2025